SCHOOL TO HOME A

McGRAW-HILL

SCIENCE

GRADE 3

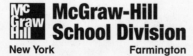 **McGraw-Hill**
School Division

New York   Farmington

McGraw-Hill School Division

A Division of The McGraw-Hill Companies

McGraw-Hill School Division
Two Penn Plaza
New York, New York 10121

Printed in the United States of America
ISBN 0-02-278181-1 / 3
3 4 5 6 7 8 9 024 04 03 02 01 00

Table of Contents

Human Body: Keeping Healthy

Science Unit

Living Things

Dear Family,

Has your child ever joyfully blown the seeds from a dandelion? Is a trip to the zoo a special memory? Our class is studying living things. In this science unit, children will explore the life cycles of plants and animals and learn how living things meet their needs. They will investigate how body parts work together to help an organism survive and will be introduced to the smallest parts living things are made of — cells.

As part of our investigation, we will send home activities for you to complete as a family. These activities use everyday objects and experiences to reinforce the scientific concepts your child is learning at school. For example, you may experiment to observe how a potato grows "eyes," or you may learn how to estimate the age of a tree.

Your help can involve you directly in your child's learning. We hope that you enjoy doing these School-to-Home Activities together!

Thank you very much,

_____

SCHOOL TO HOME

The Eye of the Potato

A *tuber* is an underground stem that contains the plant's stored food. A potato is a tuber. Potatoes have buds or "eyes." The eyes have all that is needed for reproduction. Each eye can grow into a new potato plant.

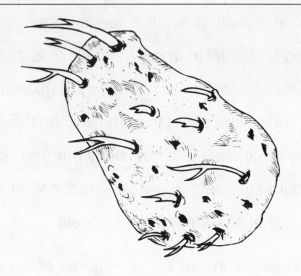

Materials

- large potato
- plastic knife
- 2 cups

What to do

1. Have an adult family member cut 2 sections from the potato. One section should contain a potato "eye," the other section should not.

2. Place the potato section with the eye in a cup. Fill the cup with 1 inch of water.

3. Do the same with the section of potato without the eye.

4. Let both cups stand for 4 days. Add water if needed.

5. Check the cups at the end of 4 days. Which potato has a sprout?

_____

Growing Up

All living things grow and change. How have you grown and changed? You grow too slowly to actually watch yourself grow, but you have probably seen pictures of when you were younger. Besides growing taller, what else about you has grown or changed? Can you predict how you will grow and change in the future?

Materials

- paper
- pencil

What to do

1. Talk with a family member about what you were like as a baby.

2. Fold your paper so there are three sections.

3. Draw a picture of how you looked as a baby in the first section.

4. Draw yourself now in the second section.

5. Finally, draw how you might look when you grow up in the last section.

6. Put your name on the back of your paper and let your teacher hang it up in the classroom. See if other children can guess who you are.

Can't Live Without 'Em

People could not live without plants. Plants provide us with many things.

You and a family member can make lists of all the items you eat or use in other ways that come from plants.

Materials

- 2 sheets of paper
- 2 pencils

What to do

1. On each sheet of paper, make 2 columns. Label the first column "Food from Plants." Label the second column "Other Items From Plants."

2. Both you and a family member should fill as many items in each column as you can think of.

3. When you are finished, trade lists.

4. Who was able to list more items?

5. Discuss with your family member the importance of plants in everyday life.

Food from Plants	Other Items from Plants

My Life as a Vegetable

What's your favorite vegetable? Imagine you are a seed of that vegetable. Think about all the growth stages you'll go through from being a little seed to becoming a delicious, ripe vegetable.

Materials

- paper
- pencil

What to do

1. Write a diary of your life as a vegetable. Start as a seed and go through the growth stages. Make an entry for each stage. Tell what the sun, soil, rain and wind are doing to you. What sounds might you hear?

2. Ask a family member to read your diary.

3. Rewrite each diary entry on one small piece of paper. Make a cover for your diary. Staple all the pages together at the left-hand side.

Rings Around the Tree

There are two ways to find out how old a tree is. One way is to look at a tree trunk that has been sawed off. Count the number of dark rings inside the trunk. Each ring represents a year's growth.

You and an adult family member can find out the age of a tree that has not been cut down.

Materials

- piece of string or cord
- tape measure or ruler

What to do

1. Have the adult family member put the string around the tree 5 feet above the ground. Mark the string where it completely circles the tree.

2. Measure the marked string with your ruler or tape measure.

3. Most trees grow about 1 inch wider each year. Count the number of inches you measured on your string. Your answer will be the approximate age of the tree.

4. Keep a record of your measurements. In one year, measure the tree again to see how much it grew.

© McGraw-Hill School Division

Science Unit

Lift It, Push It, Pull It

Dear Family,

Has your child ever wondered how fast a car can drive or why a tossed ball falls back to Earth? Our class is studying how forces like friction and gravity affect motion. Children will learn how to describe position and speed and will experiment to demonstrate how the motions of objects can change. They will explore how energy is used to do work and how simple machines like levers, pulleys, and screws can make work easier.

As part of this investigation, we will send home activities for you to complete as a family. These activities can involve you directly in your child's learning. For example, you may demonstrate how pushes and pulls work in your child's world by discussing the movements of games and sports together. In another activity, you may identify machines around the home.

Your help with these School-to-Home Activities will enhance the lessons your child is receiving in the classroom. We hope that you and your child enjoy discovering science together!

Thank you very much,

_____

Games We Play

In all of the games and sports that you play, you use pushes and pulls. These are forces that are very important, especially when you catch, run, throw, or perform other actions in games and sports. When you throw a ball in baseball, you are using a pushing force against the ball. When you play "pick up sticks," you are using a pulling force to pull out the sticks.

You and a family member are going to talk about the pushes and pulls you use when you play games.

Materials

- paper
- pencil
- crayons (or drawing pencils)

What to do

1. Make a list of your favorite sports and games. For each one, write how you use forces, and tell if they are pushes or pulls.

2. Pick one game or sport from your list and draw a picture of how you use forces when you are playing. Label these forces pushes and pulls.

© McGraw-Hill School Division

_____

_____

_____

_____

_____

Machines Around the Home

You are learning about machines in science. This home activity will give you an opportunity to see how the many kinds of machines are important.

Materials

• paper   • pencil

What to do

1. Look around the house and make a list of machines you use to do work. Some examples of machines found at home are: brooms, bottle openers, and washing machines.

_____   _____

_____   _____

_____   _____

_____   _____

2. Draw a circle around each compound machine on your list.

What did you learn?

1. How do you know if a machine is a compound or simple machine?

2. List three different ways that you could group the machines.

a. _____

b. _____

c. _____

Using what you learned

3. What safety rules should be followed with two machines at home?

_____

_____

Science Unit

Matter and Energy

Dear Family,

We are studying matter and energy in science class. Children will learn how to measure mass and volume and how to compare solids, liquids, and gases. They will be introduced to metals and elements. Finally, they will explore three different forms of energy that we rely on everyday: heat, light, and electricity.

As part of our class study, we will send home activities for you to try as a family. These activities can involve you directly in your child's learning. For example, you may observe how different colors reflect light differently or experiment with magnets to learn if an object has iron in it. You may also discuss how electricity is important to life in your home.

Your help will reinforce the science lessons your child is receiving at school. We hope that you both enjoy doing these School-to-Home Activities together!

Thank you very much,

_____

What Is Attracted?

A magnet will attract an object that has iron in it. You and a family member can find out which objects in your home a magnet will attract.

Materials

- magnet
- pencil
- 10 small objects

What to do

1. You and a family member should gather 10 small household objects.

2. Write the name of each object in the first column on the chart.

3. Take turns guessing if each object will be attracted to the magnet.

4. Then take turns placing the magnet very close to each object. Is that object attracted to the magnet? Write yes or no in the second column in the chart. Who had more correct guesses?

5. What is each object that is attracted to the magnet made of?

Name of Item	Results	What Item Is Made Of
1.		
2.		
3.		
4.		
5.		
6.		
7.		
8.		
9.		
10.		

Take a Moment to Reflect

Reflection is the bouncing back of light from a surface. White objects reflect all the colors of the visible spectrum. Black objects reflect no colors. In fact, they absorb almost all the colors of the visible spectrum. You and a family member can see and feel the amount of light and heat reflected from white and black objects.

Materials

- piece of white paper
- piece of black paper
- flashlight

What to do

1. Place the white and black papers side by side on a table.

2. Darken the room and turn on the flashlight.

3. Have your family member hold the flashlight about 6 inches above the white paper and then above the black paper.

4. Looking only at the papers, notice the amount of light in the room. When is the room brighter? When is the room darker? _____

_____

I wonder if I should wear black or white today.

5. Now touch each piece of paper. Is one warmer than the other?

_____

6. Discuss with your family member when you should wear white clothes and when you should wear dark clothes.

Name_____

Activity 10
Use with Topic 5

SCHOOL
• TO •
HOME

Page 1

What Makes Colors?

Grass looks green to us because grass reflects green light to our eyes. The grass absorbs all the other colors of the spectrum.

Black objects appear black because they absorb almost all the colors of the visible spectrum and reflect very little light to our eyes.

White objects reflect all the colors of the visible spectrum to our eyes.

Materials

- paper
- pencil

What to do

The colors from the visible spectrum are listed on the next page. Take turns with a family member filling in the colors that each object reflects to our eyes. Objects reflect the colors of the spectrum that we see. Also fill in the colors that each object absorbs. Objects absorb the colors of the spectrum that we do not see.

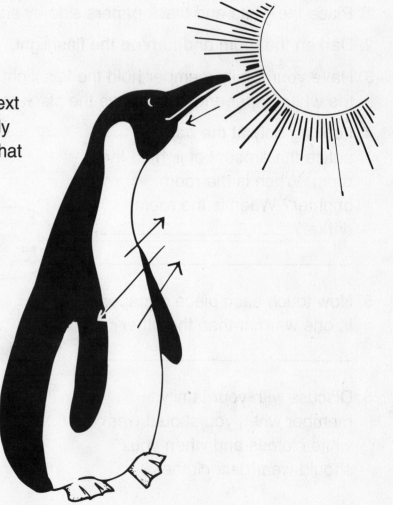

McGraw-Hill Science Unit: MATTER AND ENERGY

What Makes Colors?

Visible Spectrum						
Red	Orange	Yellow	Green	Blue	Indigo	Violet

1. Apples reflect _____

      absorb _____

2. Leaves reflect _____

      absorb _____

3. Crows reflect _____

      absorb _____

4. Bananas reflect _____

      absorb _____

5. Carrots reflect _____

      absorb _____

6. Snowflakes reflect _____

      absorb _____

7. Violets reflect _____

      absorb _____

8. Blue jeans reflect _____

      absorb _____

How Is Electricity Important to You? Page 1

You are learning about how the energy called electricity affects your life. In this activity, you will think about how you use electricity every day at home.

Materials

- paper
- pencil

What to do

1. Make a list of all the ways you use electricity at home.

_____

_____

_____

_____

_____

_____

_____

_____

_____

_____

_____

_____

How Is Electricity Important to You? Page 2

2. Look at your list. Check each one you could live without.

3. Draw a circle around those you could not live without.

4. Write a paragraph. Tell what you would do if you did not have electricity for one day. Be sure to tell what you could not do.

_____

_____

_____

_____

_____

_____

_____

_____

What did you learn?

5. How important is electricity to our daily lives at home and school.

_____

_____

_____

_____

_____

_____

_____

On your own

Make a mobile showing uses of electricity. Use coat hangers, wire or string, cardboard, and pictures to make it. Hang it in your home.

Name_____

Science Unit

The Sun and Its Family

Dear Family,

Do the stars and planets fascinate your child? We are all in
awe of the majestic expanse of outer space and the mystery of
its vastness. In science, our class is studying the Sun and the
planets of our solar system. Children will learn how the
relationship between the Sun and Earth cause night, day, and
seasons. They will explore the shape and surface of the moon.
Finally they will investigate the inner and outer planets.

As part of our class study, we plan to send home activities for
you to work on as a family. These activities can involve you
directly in your child's learning. For example, you may keep a
diary of the night sky together or write thank you notes to the
Sun. You may create a model of your own solar system or play
a game to practice the name of the planets in our solar system.

Your help will reinforce the lessons your child is receiving at
school. We hope that you and your child enjoy these School-to-
Home Activities!

Thank you very much,

_____

Sky Diary

How often do you look up into the night sky? You and an adult family member can start a sky diary together.

Materials

- pencil
- notebook

What to do

1. You will make a Moon diary and a star diary.

2. Starting tonight and for the next 5 nights, go outside after dark with your family member and look up at the sky.

3. Draw the night sky on pages of a notebook. If the sky is cloudy on any night, note that in the diary.

4. Talk with your family member about the Moon phases and also about star size, color, and brightness.

5. Draw the Moon phases that you see in the night sky. Write the Moon's phase.

6. Draw the patterns of the stars that you see.

Name_____

Activity 12
Use with Topic 2

Page 2

SCHOOL
• TO •
HOME

Sky Diary

The Moon is Earth's natural satellite. The Moon revolves around Earth every 27 days. As it revolves it spins on its axis. Moon phases are the changes in the Moon's shape as seen from Earth. We can't see a **new Moon** because the lighted side is facing away from the Earth. A new Moon happens when the Moon is between Earth and the Sun. We say the Moon is **waxing** as we see more of its lighted side each night. We can see the completely lighted side of a full Moon. A **full Moon** happens when Earth is between the Moon and the Sun. We say the moon is **waning** as we see less of its lighted side.

What to do

7. For your Moon diary, draw the Moon phases that you see in the night sky.

8. Next to each of your Moon pictures, write the Moon phase.

Moon Phase

Night 1

_____

Night 2

_____

Night 3

_____

Night 4

_____

Night 5

_____

Sky Diary

The Sun is the star in our solar system. It is so bright during the day that we can't see any other stars in the sky. At night, after the Sun goes down, we can see other stars.

What to do

Go outside with your adult family member after dark. Look at the night sky. Do you see patterns or groups of stars?

For your star diary, observe a different portion of the sky on each of 5 different nights. Draw the star patterns that you see each night.

Did you and your family member notice that some stars are brighter than others? Could you see that some are even different colors?

Night 1

Night 2

Night 3

Night 4

Night 5

Thank You, Sun!

All life on Earth depends on the energy from the Sun. How many things can you think of to thank the Sun for?

Materials

- 2 sheets of paper
- 2 pencils

What to do

1. Make thank-you lists for the Sun. Put anything on your list that you can think of that comes from the Sun. Ask a family member to do the same.

2. Trade lists. Whose list is longer?

3. Talk about your lists with each other.

Your Own Solar System

In our solar system, there are 9 planets that revolve around the big star called the Sun. Each planet has its own name. Each planet is a different size. Each one is a different distance from the Sun. Some planets are made of gases, some are made of rocks. Some planets have rings and some have moons. In this activity you and a family member have discovered a new solar system somewhere in space.

Materials

- 2 pencils
- 4 or 5 sheets of paper

What to do

Use our own solar system as a guide.

1. Make up names for the big star (the Sun of this solar system) and the planets that revolve around it.

2. Decide how far away the planets are from the big star.

3. Decide if life exists on them. What kind of life? What are they made of?

4. Draw a picture of your make-believe solar system.

5. Make a table that includes the names and sizes of the planets, as well as how far from the big star they are.

6. Write a short story about how your solar system began.

Solar System Sentences

Nine planets revolve around the Sun in our solar system. In order, starting from the Sun, these 9 planets are Mercury, Venus, Earth, Mars, Jupiter, Saturn, Uranus, Neptune, and Pluto. If you take the first letter of each planet you get MVEMJSUNP.

What to do

1. Sometimes it is fun to make up a silly sentence to help you remember something.

2. Each word in the sentence starts with one of these letters, going in order. This will help you remember the order of the planets.

3. Make up some sentences with a family member. See who can come up with the funniest sentence.

My vegetables eat mostly jelly sandwiches until nine P.M.

Science Unit

Rocks and Resources

Dear Family,

Has your child ever asked you how a volcano works or where diamonds come from? In science class, we are studying Earth's rocks and resources. Children will learn how slow processes like weathering shape rocks and how fast processes like volcanic eruptions dramatically change the land around us. They will discover the sources of natural resources like gems, minerals, oil, and gas and consider how Earth's precious resources can be conserved.

As part of our investigation, we will send home activities for you to try as a family. These activities use everyday objects and experiences to reinforce the scientific concepts we are working on at school. For example, you may compare new and old shoes to demonstrate how objects can be worn down with use over time. In other activities you may brainstorm to find solutions to pollution in your area or play a game to practice new science vocabulary.

We hope you will enjoy doing these School-to-Home Activities together!

Thank you very much,

_____

Something Old, Something New

Did your family ever have an old sneaker contest? Some sneakers really get worn out. In science class you have been learning that even rocks wear out. In this activity you can compare worn-out shoes with weathering rocks.

Materials

• several pairs of shoes, some old, some new

What to do

1. Together, look at the soles of an old pair of shoes carefully. Do you notice any patterns in the way they were worn out?

2. What kind of surface did those shoes walk on? Was it on a road, inside buildings, or out on a grassy playing field?

3. Does the surface make a difference in the way the shoe wears out?

4. Compare the old shoe with a newer shoe of the same type. What is the difference?

What did you learn?

1. What were some of the things that made parts of the shoe wear out?

_____

_____

2. What are some of the things that cause rocks to wear away?

_____

_____

3. Which of these things are the same?

_____

_____

Gold Fever

Ores are rocks containing valuable minerals. In this game, you and a family member can prospect for valuable ores.

Materials

- pencil
- paper clip

What to do

1. Make a spinner with the pencil and paper clip as shown.

2. Take turns spinning the paper clip.

3. Keep score on the sheet of paper.

4. Each of you should take 10 turns.

5. The winner will have more points at the end of 10 turns.

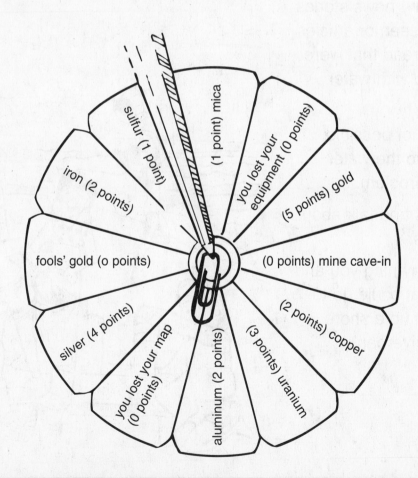

Pollution Solutions

Pollution is a big problem in our oceans, lakes and rivers. Water pollution affects all the living resources in a body of water. Pollution problems such as oil spills, garbage, and waste water harm many living organisms.

You and a family member can discuss water pollution and think of possible pollution solutions.

Materials

- 2 sheets of paper
- 2 pencils

What to do

1. Discuss any news stories you have seen or articles you have read that were concerned with water pollution.

2. Make lists of possible solutions to the water pollution problem.

3. Trade lists and talk about your ideas.

4. Is there anything you and your parent could do to help clean up a shore beach or riverbank nearby?

© McGraw-Hill School Division

Science Unit

Where Living Things Live

Dear Family,

Has your child ever discovered a bird's nest near your house? Where was it located? Was its shape or its contents difficult to see? In science class, we are studying the places where living things live. Children will learn how the special characteristics of plants and animals help them to live in different environments. They will study how organisms occupy roles in food chains and how these roles create relationships of cooperation and competition within communities. Finally children will consider how entire ecosystems can change over time.

As part of our class study, we will send home activities for you to try as a family. These School-to-Home Activities can involve you directly in your child's learning. For example, you may go on a walk to explore what living things make their homes near yours, or you may discuss how the people that share your home work together to meet your family's needs.

By sharing the questions and responsibilities in the activities with your child, you will enhance the science instruction your child is receiving at school. We hope that you enjoy discovering the natural world together!

Thank you very much,

_____

Bird-Watching Checklist

Ask an adult family member to help you to do some bird-watching.

Materials

- pencil
- notebook

What to do

Use the bird-watching checklist found below. Choose one bird and record as many observations as you can. Describe and compare your observations with other students in your class.

Checklist

1. Where was the bird? Circle your answer(s).

ground   tree   shrub   water   air   field   marsh   suburb   city

Other? _____

2. What activities did you observe?

walking   hopping   running   flapping flight   gliding flight

wading   swimming   sitting   sleeping   feeding

Other? _____

3. Did the bird call or sing?   Yes   No

4. What color(s) was the bird? _____

5. What body shape did the bird have? Draw your answer on the back of this paper.

6. What color patterns did the bird's body have? Draw your answer on the back.

7. What did the bird eat? _____

8. Review your checklist. Use a field guide if you have one. Do you know the name of the bird you observed?

McGraw-Hill Science Unit: WHERE LIVING THINGS LIVE

© McGraw-Hill School Division

What Animals Live Near You?

Materials

- pencil

What to do

Draw any animals, tracks, or animal food that you find near a tree, cracks in the sidewalk, a bush, and the grass.

What did you learn?

1. What animals live near your home?

_____

2. To what groups do these animals belong?

_____

3. How did you find out where animals live?

_____

tree	**sidewalk cracks**
bush	**grass**

House Space

"Who shares the space in our home?" You are learning about how living things work together to meet their needs. Each person who lives in your household has a role in the survival of the family.

Materials pencil

What to do

1. Draw each person who lives with you in the house shown below and write in each of their names.

2. Now add your pets.

3. Count the number of people and animals that share the space.

4. Talk about the roles of each person who lives with you with a member of your family.

5. What does your household need to survive? _____

_____

6. What kinds of chores do you do to help your family meet its needs? _____

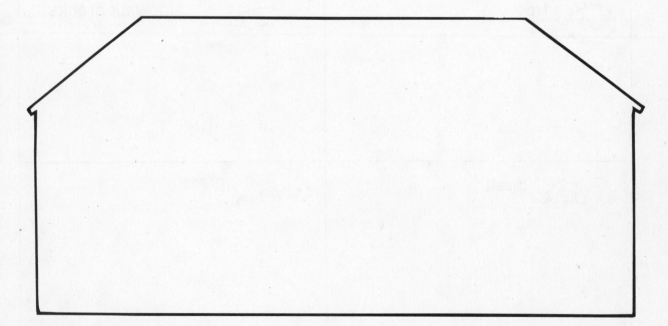

Science Unit

Human Body: Keeping Healthy

Dear Family,

Our class is studying two body systems that keep us healthy: the immune system and the digestive system. Children will investigate how the skin acts as a barrier against germs and how the body's internal defenses react to bacteria and viruses. They will learn about the different kinds of food the body needs to stay healthy and how the food we eat is transformed into usable energy.

As part of our investigation, we will send home activities for you to complete as a family. By sharing the questions in the activities with your child, you will reinforce classroom lessons and become directly involved in your child's learning. For example, you may experiment to observe how skin works to regulate body temperature, or you may investigate the nutrients in the foods you eat every day. In other activities, you may play games to practice new science vocabulary.

We hope that you and your child will enjoy doing these School-to-Home Activities together!

Thank you very much,

_____

Goose Bumps

Did you know that skin is the largest organ of your body? Among other things, skin helps to keep your body temperature even. You and a family member can see how this works.

Materials

- water faucet
- small fan

What to do

1. Wet your arm by holding it under a stream of cool water.

2. Have an adult family member position the fan so that it blows directly on your wet arm.

3. After a minute or so—the time will depend on how cool it is in your home—have your family member observe your arm. What are the little bumps called?

4. Goose bumps form due to the closing of the blood vessels in your skin. The pores in your skin also close. These reactions prevent heat from leaving your body.

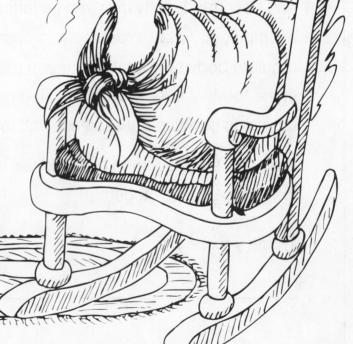

brr!

© McGraw-Hill School Division

Be a Clean Machine

Personal cleanliness is very important for good health. You and a family member can play the Cleanliness Game.

Materials

- one die (from a pair of dice)

What to do

1. Take turns. Roll the die.

2. Read out loud the Clean Fact that matches the number on the die.

3. Place a check mark by that number on the Cleanliness Game Card.

4. The first person to check off all 6 numbers is the winner!

Clean Facts

1. You bathe or shower regularly with soap. This is important for clean, healthy skin.

2. You wash your face at least twice a day with soap and water.

3. You wash your hands before each meal and snack.

4. You keep your fingernails and toenails clean and trimmed neatly.

5. You use shampoo to wash your hair at least twice a week.

6. You brush your teeth twice a day and floss your teeth often.

Cleanliness Game Card	
Me	**You**
1.	1.
2.	2.
3.	3.
4.	4.
5.	5.
6.	6.

Needed Nutrients

You need food for energy. You also need food for growth and repair of body tissues. Healthful foods contain nutrients that meet the needs of the body.

What to do

You and a family member can find out more about nutrients as you complete this puzzle.

Across

2. You need this mineral for healthy bones.

4. These are nutrients your body needs for growth and repair of body cells.

5. This keeps your body temperature normal and carries waste material out of your body.

7. Your body needs 16 kinds of these.

8. These are carbohydrates that have to be broken down into sugars during digestion before they are used by the body. Pasta and potatoes are good sources of them.

Down

1. These are your body's main source of energy.

3. These nutrients help your body use protein, fat, and carbohydrates.

4. This is a good source of protein.

6. Your body gets the most concentrated form of energy from these.

Words to Use	
calcium	proteins
carbohydrates	starches
fats	vitamins
minerals	water
poultry	

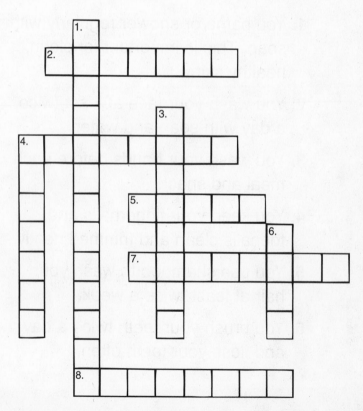

McGraw-Hill Science Unit: HUMAN BODY: KEEPING HEALTHY

Nutrient Search

Nutrients are the substances in foods that the body uses to maintain good health. Many labels on food products contain a list of the nutrients and their amounts.

Look at this nutrient label from a box of cereal. Notice that this cereal has little protein and fat but a large amount of carbohydrate.

	28.35 g serving
Protein	2g
Carbohydrate	22g
Fat	1g

What to do

Check the labels on these food products and record the amount of protein, carbohydrate, and fat per serving.

1. White Bread

Protein _____

Carbohydrate _____

Fat _____

2. Margarine

Protein _____

Carbohydrate _____

Fat _____

3. Tuna in Spring Water (can)

Protein _____

Carbohydrate _____

Fat _____

4. Fruit Cocktail

Protein _____

Carbohydrate _____

Fat _____

5. Which food is high in protein per serving? _____

6. Which foods are high in carbohydrates per serving?

_____

7. Which food is high in fat per serving?

_____

Answers

Page 2: The Eye of the Potato

5. The section with the potato "eye" should have a sprout.

Page 3: Growing Up

Examples of growth and change include height, weight, and presence of teeth and hair.

Page 4: Can't Live Without 'Em

Foods from Plants	Other Items from Plants
Examples: fruits, vegetables, grains	Examples: woods, charcoal, turpentine, rope, dyes, cotton, spices, medicines

Page 5: My Life as a Vegetable

Diaries should include the different stages of the plant growth cycle: planting of the seed, germination in the soil, growth of roots, emergence from the soil, growth of first leaves, growth of secondary leaves, flowering, fruit development, and harvest.

Page 6: Rings Around the Tree

Students might ask family members if they have ever planted any trees. Students might plant a tree now and watch it grow as they grow.

Page 8: Games We Play

Games and Sports	Forces
Baseball	Pushes
Tennis	Pushes
Golf	Pushes
Soccer	Pushes
Tug of war	Pulls
Fishing	Pulls
Roping	Pulls
Rowing	Pulls

Page 9: Machines Around the Home

Examples of simple machines:
Rake
Shovel
Hammer
Screwdriver
Wrench
Fork, knife, spoon

Examples of compound machines:
Sewing machine
Bicycle

Electric can opener
Dishwasher
Lawn mower
Car

What did you learn?

1. A simple machine has few or no moving parts. Compound machines are made of two or more simple machines put together.

2. Machines may be grouped by type of use, by number of moving parts, by size, or by materials they are made from.

Using what you learned:

1. Answers will depend on the machines chosen but may include being careful with electricity, heat, and moving parts.

Page 12: What is Attracted?

5. Objects with iron or steel are attracted to magnets.

Page 13: Take a Moment to Reflect

4. The room is lighter when the light is shone on the white paper and darker when the light is shone on the black paper.

5. The black paper is warmer.

6. White clothes keep our bodies cooler in the summer; dark clothes keep our bodies warmer in the winter.

Pages 14–15: What Makes Colors?

1. reflect: red
absorb: orange, yellow, green, blue, indigo, violet

2. reflect: green
absorb: red, orange, yellow, blue, indigo, violet

3. reflect: none
absorb: red, orange, yellow, green, blue, indigo, violet

4. reflect: yellow
absorb: red, orange, green, blue, indigo, violet

5. reflect: orange
absorb: red, yellow, green, blue, indigo, violet

6. reflect: red, orange, yellow, green, blue, indigo, violet
absorb: none

7. reflect: violet
absorb red, orange, yellow, green, blue, indigo

8. reflect: blue
absorb: red, orange, yellow, green, indigo, violet

Pages 16–17: How Is Electricity Important to You?

What to do:

1. Electricity is used for heating, lighting, cooking, and powering appliances such as radios, televisions, computers, and microwave ovens.

What did you learn?

5. Include energy conservation in the discussion.

Pages 20–22: Sky Diary

8. There are eight moon phases: new moon, waxing crescent, first quarter, waxing gibbous, full Moon, waning gibbous, last quarter, waning crescent.

Gibbous means rounded. Waning means decreasing. Waxing means increasing.

9. A star's brightness is a function of both size and distance. A star's color, which can sometimes appear as white, yellow, red, or blue to the naked eye, is a function of size and age.

Page 23: Thank You, Sun!

Answers may include giving thanks for light, warmth, plants, summer, beaches, flowers, and shadows.

Page 24: Your Own Solar System

Systems and stories will reflect the students' imaginations but may include suns, small planets, large planets, distances from suns, oceans, life, icy or barren landscapes, gaseous atmospheres, craters, rings, asteroids, and moons.

Page 25: Solar System Sentences

Example: My very educated mother just served us nine pizzas.

Page 28: Something Old, Something New

What to do:

1. Yes, there are patterns for how soles, toes, or sides of shoes wear out.

2. Answers may include grass, pavement, tile, or linoleum.

3. Yes, surface affects how quickly and in what way shoes wear out. Walking on concrete pavement will wear shoes out more quickly than walking on other surfaces.

4. The old shoe is worn away in places.

What did you learn?

1. Examples may include friction with road pavement, dirt, beach sand, cement, water, or ice.

2. Examples may include friction with running water, ice, wind, sand, and other rocks.

3. Ways that both shoes and rocks wear away include friction with sand and water.

Page 29: Gold Fever

As an extension of the game, students might research different kinds of ores.

Page 30: Pollution Solutions

2. Possible solutions to water pollution may include care with the location of land dumping, cleaning up water front areas, refining water purification processes, and refraining from building near and dumping in water environments.

3. Contact a local or national environmental or marine protection organization for clean-up programs and dates.

Page 32: Bird-Watching Checklist

All answers depend on the family's observations.

Page 33: What Animals Live Near You?

What did you learn?

1. Examples include insects, squirrels, birds, skunks, and pets.

Examples of groups include insects, mammals, rodents, birds, reptiles, amphibians.

3. being still, using eyes and ears to observe

Page 34: House Space

To survive, the members of a household need a shelter, heat, water, and food. Student's answers may also include the need for income, jobs, transportation, electricity, and furniture.

Page 36: Goose Bumps

3. The little bumps on the skin that form when you are chilled are known as goosebumps. When the hairs on your skin stand up, your body is making a layer of still air around its surface to reduce the cooling effect of breezes.

Page 37: Be a Clean Machine!

Students might discuss how they help keep their homes and classrooms clean.

Page 38: Needed Nutrients

Across:

2. calcium
4. proteins
5. water
7. minerals
8. starches

Down:

1. carbohydrates
3. vitamins
4. poultry
6. fats

Page 39: Nutrient Search

Nutritional information for typical examples of food products:

1. bread:
   Protein: 2g; Carbohydrate: 13g; Fat: 1.5g

2. margarine:
   Protein: 0g; Carbohydrate: 0g; Fat: 11g

3. tuna:
   Protein: 13g; Carbohydrate: 0g; Fat: 0.5g

4. fruit:
   Protein: 0g; Carbohydrate: 20g; Fat: 0g

5 tuna

6. bread, fruit

7. margarine